Sparks pop. Lightning strikes. City lights glow,
washing machines spin, and refr
are just a few of the visible effec
invisible form of energy known a

Electricity itself is invisible because it has to do with the behavior of tiny bits of matter called *electrons,* which whirl around the cores of atoms. An atom's electrons and core carry opposite electrical charges. Scientists call the charges that electrons carry *negative* charges. They call the charges that the cores carry *positive* charges.

Static Electricity

Have you ever brushed your hair and heard a slight crackle? Touched a doorknob, seen a little spark, and felt a mild shock? If you have, then you've experienced static electricity—a buildup of electrical charges that can also make some objects attract and repel one another like magnets.

Ordinarily, a plastic comb won't pick up little pieces of paper like this one does. But if you rub a comb, ruler, or other plastic object with a piece of cloth, you can create a buildup of charges by stirring up electrons.

As you rub, some electrons leave atoms in the cloth and move to atoms on the surface of the plastic. The cloth becomes *positively charged*—that means it has lost electrons. The plastic becomes *negatively charged*—it has gained electrons.

To understand what happens next, you need to remember that like charges repel and opposite charges attract. As you move the plastic close to some bits of paper, the negative charges repel electrons near the surface of the paper. These electrons move out of the way, leaving the surface of the paper positively charged. Presto! Now the plastic attracts the paper.

Lightning is the result of nature's most dangerous form of static electricity. Areas of opposite charges build up in clouds. Negative charges then leap to a positively-charged spot, producing a streak of light as the air heats up and a clap of thunder as it suddenly expands.

The first known experiments with electricity were performed in ancient Greece, when a philosopher named Thales (THAY-leez) rubbed pieces of amber with fur or wool. In fact, the word *electricity* comes from *elektron,* the Greek word for amber.

Electricity in Circuits

Static electricity can cause sparks and make some objects attract others. But it can't light a light bulb or run useful devices like toasters or vacuum cleaners. To run these machines, you need *current electricity*—electricity that flows continuously. In order to flow continuously, the electricity must travel through a loop called a *circuit*.

This model diner is equipped with lights that are wired in circuits. Inside the circuits, electrons are on the move. As they jump ahead from atom to atom, they repel other electrons—which also jump, repelling and moving other electrons. We call this invisible stream of movement an *electric current*.

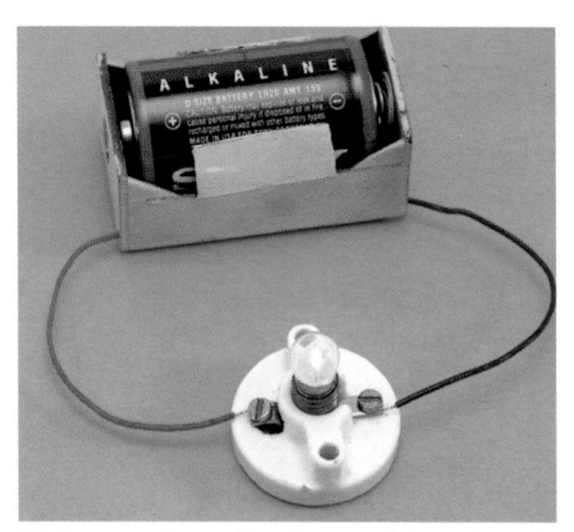

Take a close look at this simple circuit. Like all circuits, it is arranged in a loop. Electric current flows through the battery, wire, and bulb in a complete, unbroken path.

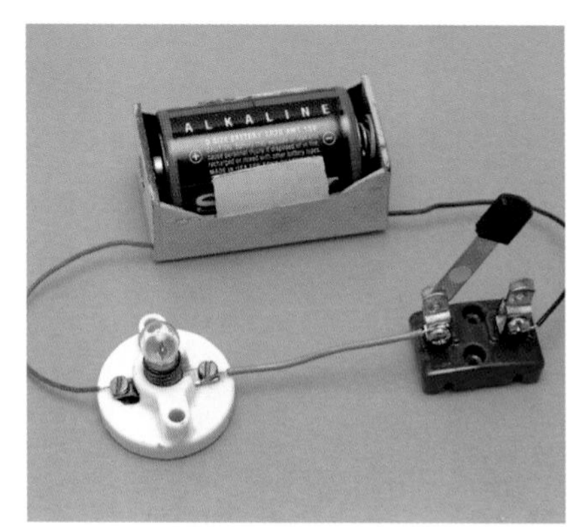

Most circuits have switches, which let you turn the current on and off. When the switch is in the "off" position, the circuit is incomplete and the current cannot flow in a complete path.

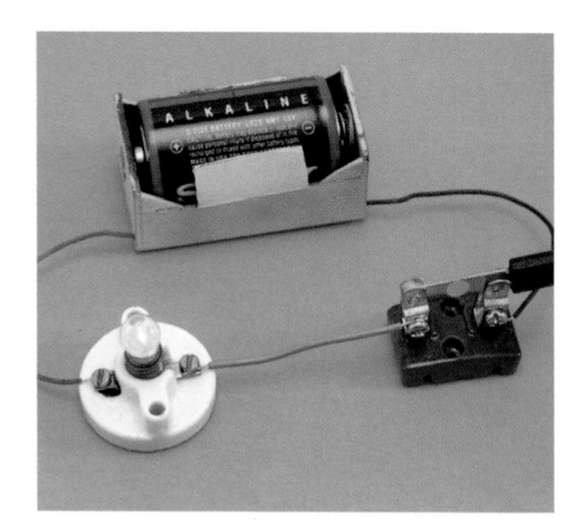

To turn the light back on, you flip the switch the other way. The circuit is now complete, and the current flows again.

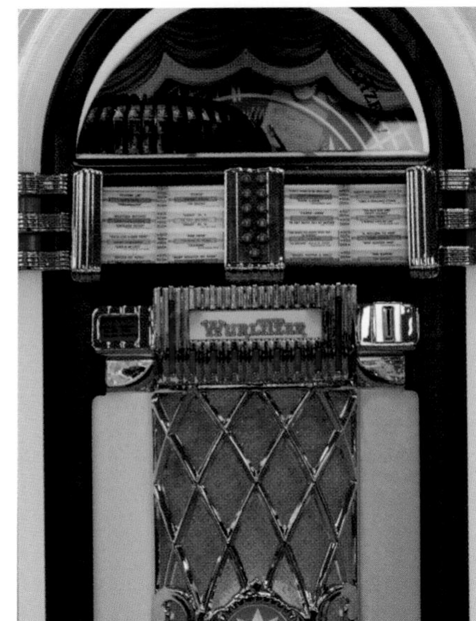

We use electrical devices to produce not only light but also heat, sound, and movement. Every time you flip a switch, call a friend, play a song on a jukebox, or heat up soup on an electric burner, you are sending electricity into action by completing a circuit.

Batteries

Every circuit needs something to push the electrons along. In a simple circuit that you make yourself, the push comes from a battery.

You can use a lemon and two pieces of different metals to make a battery. Every battery contains a liquid or paste, called an *electrolyte* (eh-LECK-troh-lite). In this battery, lemon juice is the electrolyte. The juice chemically reacts with both metals, leaving one positively charged and the other negatively charged. When the metal parts are connected in a circuit, electrons flow through the wire, pushed from the negative to the positive end.

The batteries we use every day also produce electricity by means of chemical reactions (changes in substances). But the chemicals inside these batteries are much stronger and more dangerous than lemon juice. *Never* try to break open a battery. And *never* touch a leaky battery.

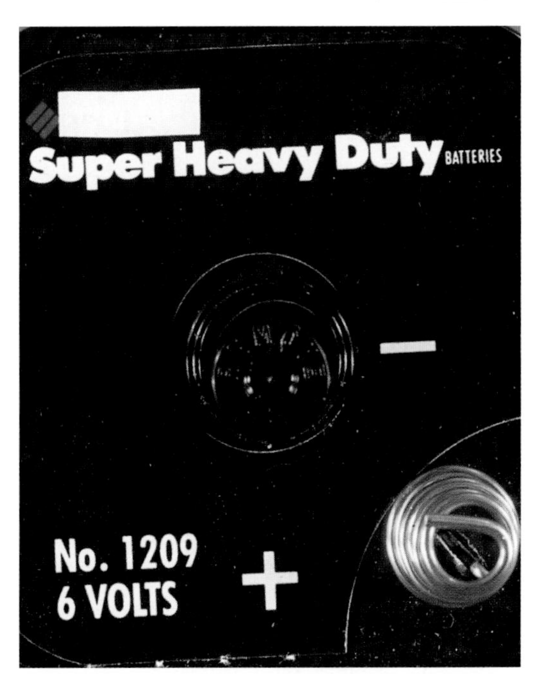

On every battery, the parts marked "plus" and "minus" tell you which way the electrons will flow. The number of volts tells you how strong a push the battery provides.

Batteries are handy and portable. They provide electricity for flashlights, watches, calculators, toys, radios, and many other small devices that you might want to carry around with you.

Conductors and Insulators

Materials that can carry an electric current are known as *conductors*. Copper, aluminum, silver, and most other metals are good conductors. Materials that cannot carry an electric current are known as *insulators*. Porcelain, plastic, and rubber are all good insulators. Both conductors and insulators play important parts in household circuits.

The atoms that make up most metals have "free electrons"—electrons that are not strongly attracted to their own atoms and so can easily hop to other atoms. The atoms that make up plastics, on the other hand, keep a tighter hold on their electrons. This means that an electric current—the movement of electrons from atom to atom—can easily pass through a metal wire but not through a piece of plastic.

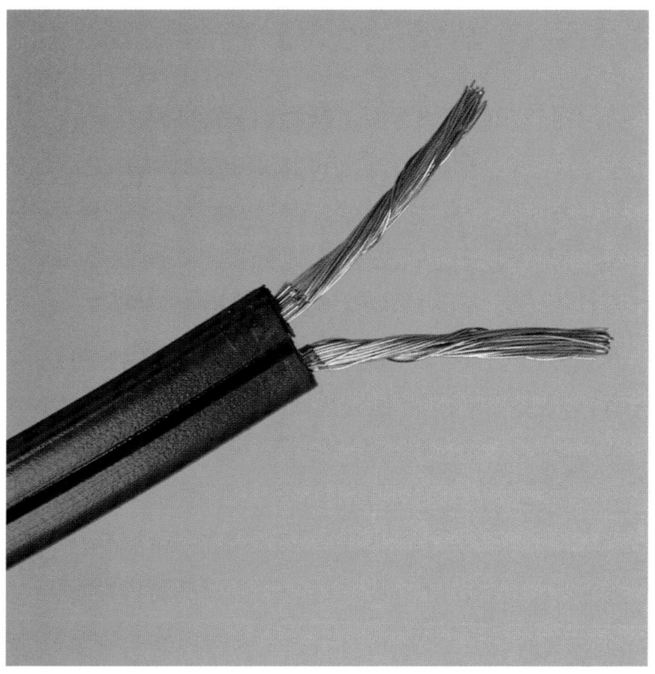

Lamps and other devices that plug into walls have electric cords. Inside the cords are wires—usually copper—that can conduct electricity. Outside, plastic coverings act as insulators.

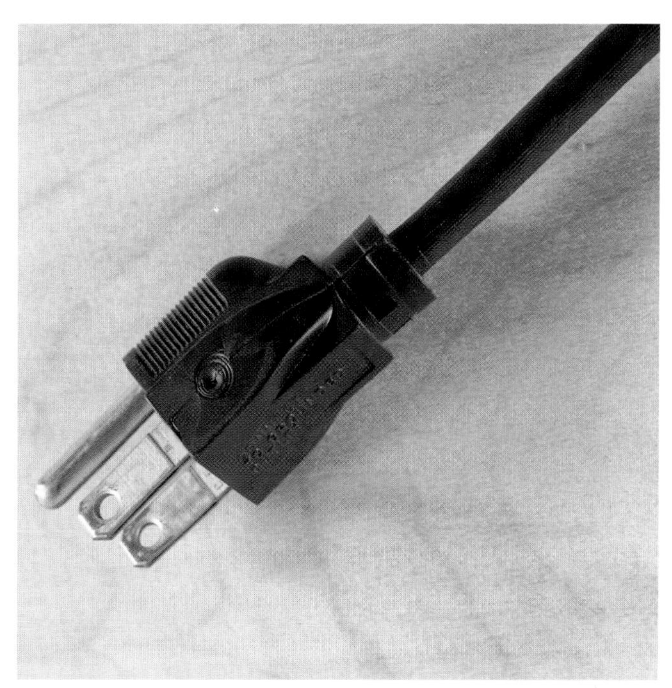

The plugs at the ends of electric cords are also made of conducting and insulating parts.

Outlets have insulated coverings. They also have metal parts that connect to wiring inside walls. When a device is plugged in and switched on, current runs through the metal parts of the wiring, outlet, plug, and cord. The insulators protect you from coming into contact with the current.

Household current can be dangerous. It is much stronger than the current you generate with a small battery. *Never* touch the metal parts of a plug when you plug in any appliance. *Never* play with electric outlets. You could get a very big shock and be badly hurt.

Electricity and Magnetism

In the early 1800s, two scientists made two different but closely related discoveries. One found that electricity could produce magnetism. The other found that magnetism could produce electricity. These discoveries about the connection between electricity and magnetism led to many inventions that changed the way people lived and worked.

Hans Christian Oersted, a Danish scientist, made a great scientific breakthrough when he discovered that a wire carrying an electric current also produces a magnetic field. You can make this discovery on your own by making an *electromagnet*.

An electromagnet—a piece of iron or steel surrounded by coils of wire—is a temporary magnet. It is magnetized only while current flows through the wire. Powerful electromagnets on cranes lift heavy loads of metal. When the electromagnets are switched off, the cranes release their loads.

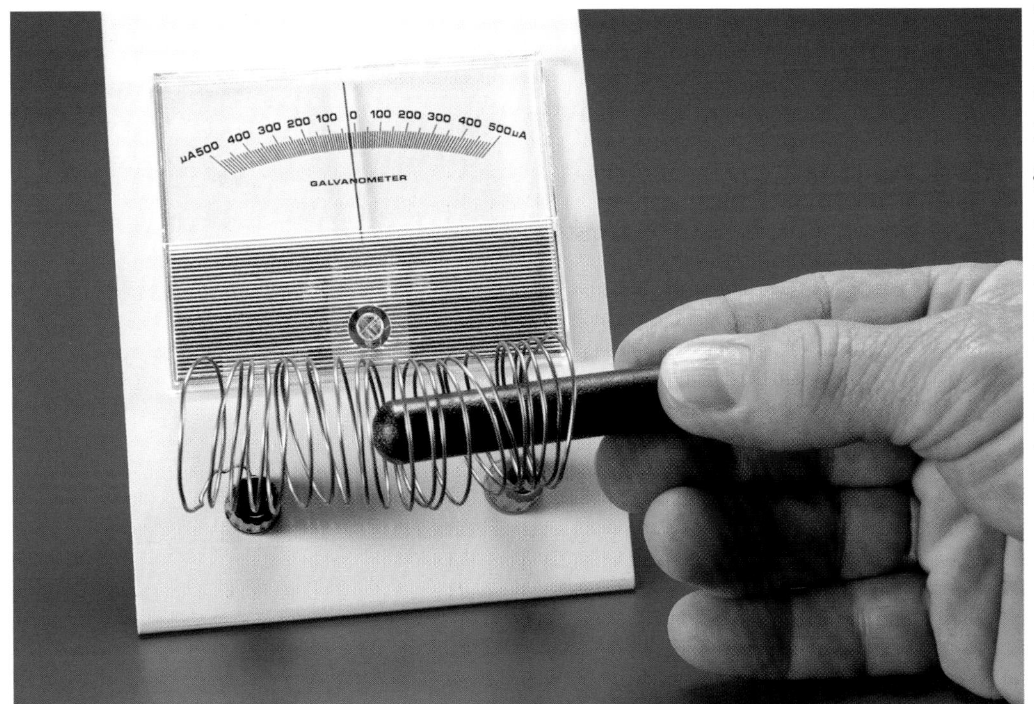

Michael Faraday, an English scientist, discovered that a magnet can make—or *generate*—electricity. You can reproduce Faraday's discovery by moving a magnet in and out of some coils of wire. What happens? As the meter shows, an electric current flows through the wire.

Faraday used his findings to invent the electric *generator*—a machine that combines magnets, coils of wire, and movement to make electricity. He also invented the first electric *motor*. Electric motors use the attracting and repelling forces of magnets and electromagnets to drive moving parts.

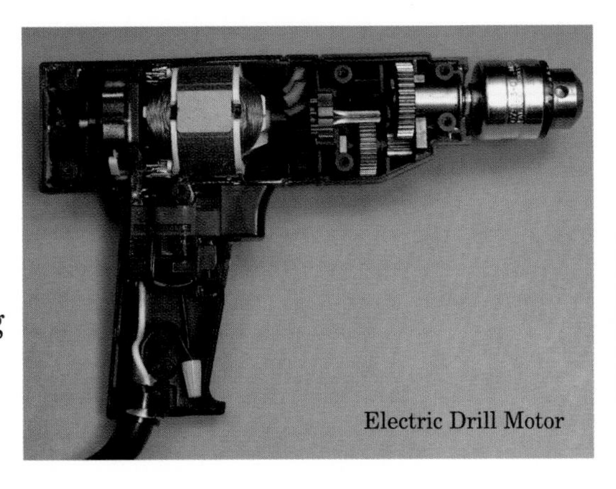
Electric Drill Motor

Telephones, televisions, tape players, CD players, and computers all use electricity and magnetism to produce sound, pictures, and movement.

Power Stations

At a power station, giant magnets spin to produce electricity. From this source, the electricity is carried along power lines in the biggest circuit of all—the one that delivers electrical power to your home, your school, and other buildings.

1 Most power stations burn coal, oil, or natural gas to boil large amounts of water. The boiling water produces steam. The pressure of the steam turns a device inside the station called a *turbine*.

2 The turbine turns a magnet—which is surrounded by a large conducting coil—inside a generator. As the magnet turns, electric current is produced in the coil.

3 Electricity travels better over long distances if its voltage—the measure of how hard it pushes—is high. So the voltage is increased, or "stepped up," in a device called a *transformer*.

4 High-voltage lines carry the electricity.

5 Closer to home, another transformer decreases, or "steps down," the voltage.

6 Low-voltage lines carry the stepped-down electricity. The lines may be strung on poles or buried under the ground.

7 Small transformers attached to the low-voltage lines step the voltage down again. Service lines feed the electricity into your home.

Electricity and the Environment

Burning coal, oil, or gas causes air pollution. Also, our supplies of these fuels will probably run out some day. That's why it's important to conserve energy and use less electricity. And that's why scientists are developing other ways to produce electricity.

Hydroelectric power stations use falling water to produce electricity. At a hydroelectric station, water from a river is held back in a reservoir. Some of the water is sent into big pipes to turn turbines, which turn generators. Stations like these now supply about one-fifth of the world's electricity. They do not cause air pollution, but they can be built only in places that have the right kind of land and rivers.

Wind power (left) can also be used to drive turbines and generators. Energy from sunlight can be collected to heat water for steam-driven turbines, or to produce an electric current in devices known as *solar cells* (right). Both of these alternative energy sources hold a lot of promise for the future.

Here's a bright idea for saving electricity: turn off the lights when you don't need them. Remember, most of today's power stations burn coal, oil, or gas. So reducing our use of electricity means reducing air pollution.

Energy-efficient devices—devices that use less energy to do the work they are designed to do—help, too. Compact fluorescent light bulbs use less electricity than standard light bulbs, which are known as incandescent bulbs.

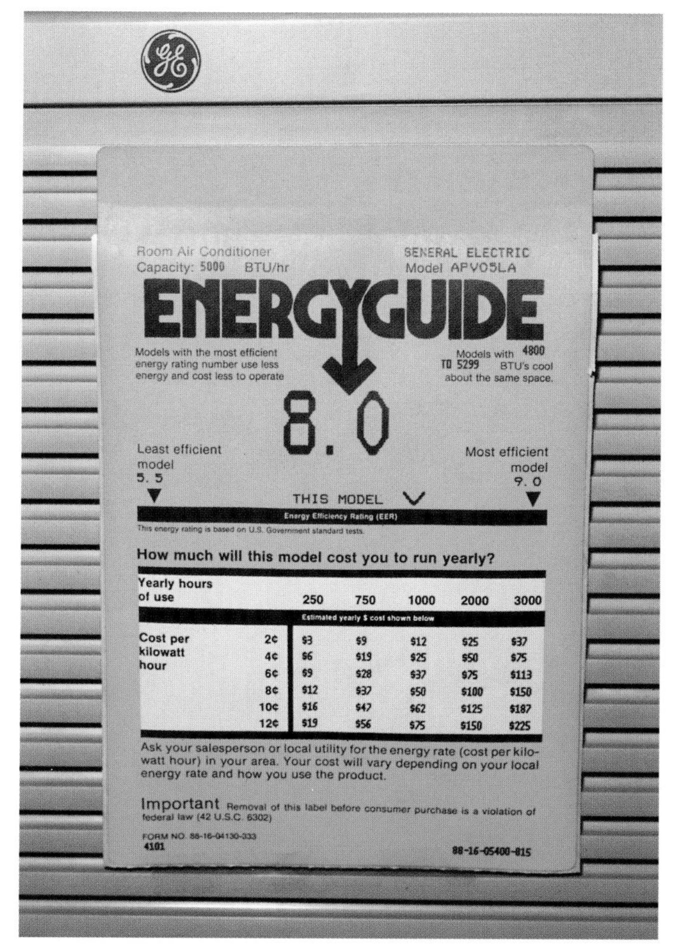

Air conditioners, refrigerators, and other big appliances are big energy users. Energy-efficient designs can save a lot of electricity.

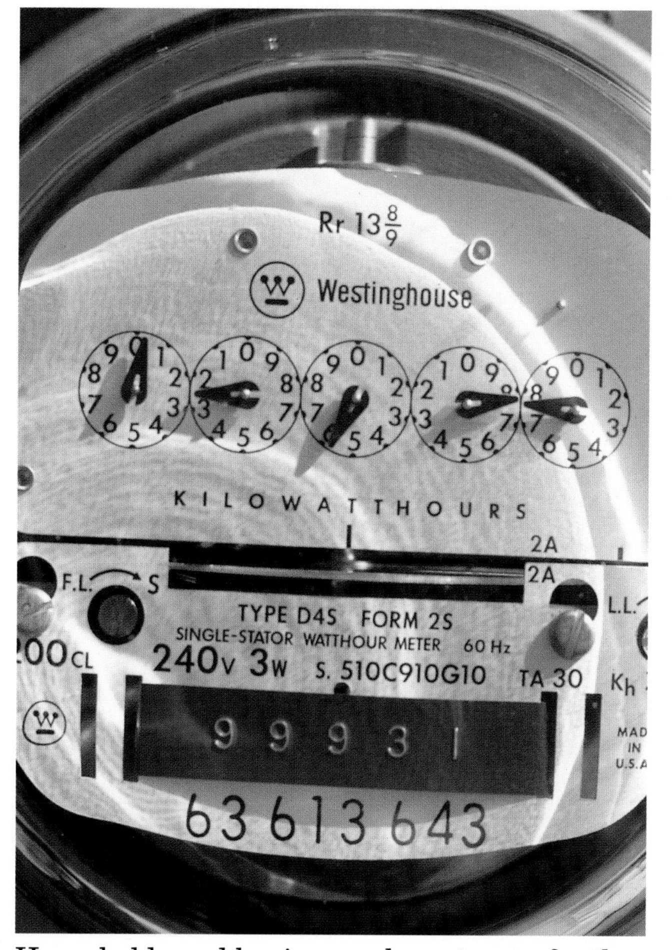

Households and businesses have to pay for the electricity they use. Saving electricity means saving money, too.

Electricity is a form of energy, and energy makes things happen. You can put your own energy to work to learn more about this invisible, useful, and sometimes dangerous force. Visit a science museum. Look for books and other sources that supply "current" information. Conduct some safe experiments. Generate your own ideas and questions!

If you visit a science museum, you might find a Van de Graaff generator like the one shown here. When you put your hands on the dome, your hair becomes charged with static electricity. Strands of hair stand up and apart because they are repelled by each other's like charges.